SIEGFRIED J. BAUER

SOLAR CYCLE INFLUENCE OF GPS RANGE ERRORS FROM MESOSCALE IONOSPHERIC ANOMALIES (MSTIDS)

ÖSTERREICHISCHE AKADEMIE DER WISSENSCHAFTEN
KOMMISSION FÜR DIE WISSENSCHAFTLICHE ZUSAMMENARBEIT
MIT DIENSTSTELLEN DES BM FÜR LANDESVERTEIDIGUNG UND SPORT

PROJEKTBERICHTE
HERAUSGEGEBEN VON HANS SÜNKEL

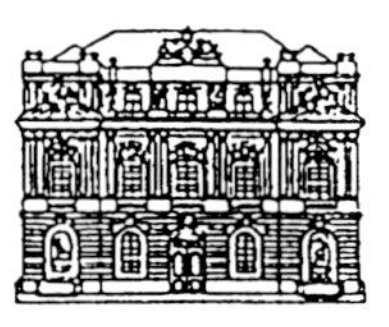

VERLAG DER ÖSTERREICHISCHEN AKADEMIE DER WISSENSCHAFTEN
WIEN 2012

SIEGFRIED J. BAUER

EINFLUSS DES SONNENZYKLUS AUF GPS-DISTANZMESSFEHLER DURCH MESOSKALIGE IONOSPHÄRISCHE ANOMALIEN (MSTIDS)

SOLAR CYCLE INFLUENCE OF GPS RANGE ERRORS FROM MESOSCALE IONOSPHERIC ANOMALIES (MSTIDS)

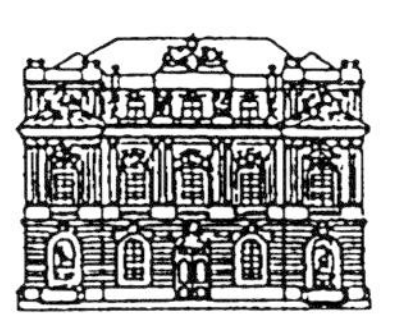

VERLAG DER ÖSTERREICHISCHEN AKADEMIE DER WISSENSCHAFTEN
WIEN 2012

Vorgelegt bei der Sitzung
der math.-nat. Klasse
am 11. Oktober 2012

ISBN 978-3-7001-7356-4

IMPRESSUM

Medieninhaber und
Herausgeber:
Österreichische
Akademie der
Wissenschaften

Kommissionsobmann:
o. Univ.-Prof. DI Dr. Hans Sünkel, w. M.

Layout:
Dr. Katja Skodacsek

Lektorat:
DDr. Josef Kohlbacher

Druck:
BMLVS/Heeresdruckzentrum
12-4251

Wien, im November 2012

Gedruckt mit Unterstützung des
Bundesministeriums für
Landesverteidigung und Sport

Editorial

Die Kommission der Österreichischen Akademie der Wissenschaften für die wissenschaftliche Zusammenarbeit mit Dienststellen des Bundesministeriums für Landesverteidigung und Sport wurde auf Initiative von Herrn Altpräsidenten em. o. Univ.-Prof. Dr. Dr. h. c. Otto Hittmair und Herrn General i. R. Erich Eder in der Gesamtsitzung der Österreichischen Akademie der Wissenschaften am 4. März 1994 gegründet.

Entsprechend dem Übereinkommen zwischen der Österreichischen Akademie der Wissenschaften und dem Bundesministerium für Landesverteidigung und Sport besteht die Zielsetzung der Kommission darin, für Projekte der Grundlagenforschung von Mitgliedern der Österreichischen Akademie der Wissenschaften, deren Fragestellungen auch für das Bundesministerium für Landesverteidigung und Sport eine gewisse Relevanz besitzen, die finanzielle Unterstützung des Bundesministeriums zu gewinnen. Von Seiten des Bundesministeriums für Landesverteidigung und Sport wird andererseits die Möglichkeit wahrgenommen, den im eigenen Bereich nicht abgedeckten Forschungsbedarf an Mitglieder der höchstrangigen wissenschaftlichen Institution Österreichs vergeben zu können.

In der Sitzung der Kommission am 16. Oktober 1998 wurde der einstimmige Beschluss gefasst, eine Publikationsreihe zu eröffnen, in der wichtige Ergebnisse von Forschungsprojekten in Form von Booklets dargestellt werden.

Meiner Vorgängerin in der Funktion des Kommissionsobmanns, Frau em. o. Univ.-Prof. Dr. DDr. h. c. Elisabeth Lichtenberger, sind die Realisierung und die moderne, zweckmäßige Gestaltung dieser Publikationsreihe zu verdanken.

Das Bundesministerium für Landesverteidigung und Sport hat dankenswerterweise die Finanzierung der Projektberichte übernommen, welche im Verlag der Österreichischen Akademie der Wissenschaften erscheinen.

Hiermit wird

- Projektbericht 14:
  Siegfried J. Bauer: Solar Cycle Influence of GPS Range Errors from Mesoscale Ionospheric Anomalies (MSTIDs). Wien 2012.

vorgelegt.

Wien, im November 2012 Hans Sünkel

Vorwort
des Kommissionsobmanns und des Obmann-Stellvertreters

Wissenschaftlicher und technischer Fortschritt sowie die Einführung komplexer technischer Wirk-, Leit-, Unterstützungs- und Hilfssysteme globaler Reichweiten, wie dies auch für die satellitengestützte Nachrichtenübertragung, Ortung und Navigation zutrifft, vollziehen sich aus Anwendersicht recht rasch, angenommenerweise wohlverstanden und für Fachleute überschaubar, sowie, so eine gängige Annahme, technisch weiter verfeinerbar mit erheblich steigerbarem Anwendernutzen.

Das Österreichische Bundesheer (ÖBH) hat sich mit aufkommender Verfügbarkeit von globalen Satellitennavigationsdiensten für die Frage zu interessieren begonnen, welche Genauigkeit, Verlässlichkeit, vor allem aber Grenzen die Anwendung dieser Systeme besonders im Bereich hoher Präzision und Zuverlässigkeit haben und ob die dazu verfügbaren bekannten Angaben hinreichen.

Der Umstand, einen Pionier und weltweit anerkannten Fachmann zu solchen Forschungsfragen in der Person von emer. o. Univ.-Prof. Dr. Siegfried J. Bauer, w. M., seit mehr als einem Jahrzehnt in den Reihen der Kommission zu haben und in Österreich auf eines der weltweit ganz wenigen hochpräzise vermessenen, terrestrischen Zweifrequenz-GPS-Empfangsstationen-Netze aufbauen zu können, hat dem ÖBH in Zusammenarbeit mit der Österreichischen Akademie der Wissenschaften erlaubt, den beschriebenen und einigen weiteren Forschungsfragen erfolgreich nachzugehen und – wie sich nachträglich herausgestellt hat, auch durch glücklichen zeitlichen Verlauf der Beobachtungen – neue, wesentliche Einsichten zu erlangen, die über die militärische Sicht hinaus auch für zivile Präzisionsanwendungen von globalen Satellitennavigationssystemen (z. B. GALILEO, GPS, GLONASS usw.) Bedeutung haben.

Der vorliegende Bericht, knapp in englischer Sprache abgefasst, beschränkt sich auf die Wiedergabe der wesentlichen physikalischen Daten und Erkenntnisse aus der Beobachtung der interessierenden Parameter über den vollen Sonnenzyklus 23 (ca. 11 Jahre). Er ergänzt und erweitert erkenntnismäßig erheblich den im Jahr 2002 in deutscher Sprache abgefassten Projektbericht 4 dieser Kommission. Strategische Ableitungen werden in der vorliegenden Veröffentlichung nicht behandelt.

Die veröffentlichten Ergebnisse dieser sich über knapp eineinhalb Jahrzehnte erstreckenden Forschungstätigkeit wenden sich an militärische, aber auch an einschlägige zivile Interessenten und werden hoffentlich auf großes Interesse stoßen.

Graz und Wien, im November 2012

o. Univ.-Prof. Dipl.-Ing.
Dr. techn. Hans Sünkel, w. M.
Kommissionsobmann

General
Mag. Edmund Entacher, ChGStb
Stv. Kommissionsobmann

Table of Contents
Inhaltsverzeichnis

Kurzfassung

Dieser Bericht stellt eine inhaltliche Ergänzung (nach 10 Jahren) des Projektberichts 4 dar, in dem erstmals Beobachtungen mit dem österreichischen Netz von zehn GPS-Stationen von mesoskaligen Ionosphären-Anomalien berichtet wurden, die nicht auf externe Ursachen („Space Weather") zurückzuführen waren, sondern von Störungen in der unteren Atmosphäre herrühren, die durch atmosphärische Schwerewellen in die Ionosphäre gelangen. In den ersten paar Jahren nach Beginn der Beobachtungen 2000 waren solche Anomalien an etwa 25 % der Tage, mit einer Bevorzugung der Wintermonate, erkennbar, jedoch weder örtlich noch zeitlich vorhersagbar; ihre Dauer war auf ein paar Stunden begrenzt, die damit verbundenen Streckenfehler im Bereich von Dezimetern bis Metern.

Die gegenwärtige Statistik von Streckenfehlern größer als 5 Dezimeter für die gesamte Beobachtungszeit von 11 Jahren ergab jedoch ein recht unerwartetes Resultat:

Die Verteilung der Fehlerfrequenz zeigte eine eindeutige Beziehung zum letzten Sonnenfleckenzyklus 23, mit dem Ergebnis, dass schon einige Jahre vor dem absoluten Sonnenflecken-Minimum (2009) keine Ereignisse mehr zu verzeichnen waren. Da die Störungsquellen in der unteren Atmosphäre offensichtlich nicht vom Sonnenfleckenzyklus abhängen, muss die Fehlerstatistik mit der Ionosphäre verknüpft sein. Die GPS-Beobachtungen liefern den Gesamtinhalt der Elektronen in der Ionosphäre (TEC), aus denen die Anomalien und Streckenfehler abgeleitet werden und dieser bestimmt offensichtlich auch unsere Fehlerverteilung. Obwohl TEC als empfindlicher Indikator der Sonnenaktivität anerkannt ist, war ein physikalischer Grund für diese Nichtlinearität nicht offensichtlich. Der Autor hat sich daher die Aufgabe gestellt, eine Begründung dafür zu geben.

Unter Verwendung der gängigen Theorie für die ionosphärische F_2-Schicht, die den Hauptanteil von TEC ausmacht, konnte gezeigt werden, dass, obwohl die Elektronen*dichte* linear von der Sonnenaktivität (EUV-Energiefluss von 1 nm bis 100 nm) abhängt, dies jedoch für den Gesamtelektronen*inhalt* nicht gilt, denn die *Integration* der exponentiell von der Höhe abhängigen Elektronen*dichte* ergibt das *Produkt* von zwei Ionosphärenparametern: die Dichte im Maximum N_m und eine effektive Skalenhöhe $H_{eff.}$ (ein Maß für die „Dicke" der Ionosphäre), beide abhängig von der Sonnenaktivität. Damit wurde ein „Verstärkungsfaktor" von 2,4 errechnet, mit dem sich TEC um einen Faktor 5 erhöht, wenn man von der Verdopplung der solaren EUV-Intensität zwischen Minimum und Maximum des Sonnenfleckenzyklus ausgeht, was auch mit Beobachtungen übereinstimmt.

Obwohl die Streckenfehler durch mesoskalige Anomalien nur relativ gering sind, (Dezimeter bis Meter), müssten diese aber für zuverlässige Navigation, besonders für sicherheitskritische Anwendungen, berücksichtigt werden. Nachdem diese Fehlerquellen derzeit praktisch nicht vorhersagbar sind, scheinen längere statistische Untersuchungen der hier beschriebenen Art von Nutzen zu sein.

Wien, im Oktober 2012 — Siegfried J. Bauer

Abstract

With the Austrian net of 10 two-frequency GPS stations, whose location is known with high precision, mesoscale variability of the ionospheric F_2 region could be investigated that is associated with medium size travelling ionospheric disturbances (MSTIDs) that apparently are initiated by atmospheric gravity waves from sources in the lower atmosphere not associated with space weather. These anomalies had an occurrence frequency of about 25 % to less than 20 % during the period from 2000 to 2005, when solar activity was elevated, producing range errors from decimetre to meters lasting up to a few hours. The extension of our database until 2012 showed a marked decrease in occurrence to a complete absence during the period of solar minimum, as evidenced by the statistics of the appearance of GPS range errors in excess of 50 cm. Since the lack of appearance of MSTIDs causing these errors cannot be due to the source phenomena in the lower atmosphere, they must be explained by the behaviour of the ionosphere and gravity wave propagation characteristics as function of solar activity. Total Electron Content (TEC), on which the GPS observations are based, has been found highly responsive to solar activity, although the reason for it has remained largely obscure. An explanation for the non-linearity between TEC and solar EUV irradiance (EUV: Extreme Ultra Violet; Energy flux from 1 nm to 100 nm) is given here on the basis of accepted ionospheric F_2 region theory, leading to the consequence that although ionospheric electron *density* reacts linearly, total electron *content* does not, since it depends on the *product* of two ionospheric parameters dependent on solar activity, the electron density at the F_2 region peak, N_m, and an effective scale height, representing the "thickness" of the ionosphere. Accordingly, an amplification factor of 2.4 is derived, that leads to an increase by a factor of 5 in TEC between solar minimum and maximum based on a known two fold increase in solar EUV irradiance.

Solar Cycle Influence of GPS Range Errors from Mesoscale Ionospheric Anomalies (MSTIDs)

1 Observational Background

GPS observations were made since 2000, after cessation of selective availability made possible the full realisation of system capability, using the Austrian network of geodetic reference stations consisting of 10 two-frequency GPS stations, whose location is known with high precision. These allow the study of mesoscale anomalies in the ionosphere over central Europe [1]. The receiver station network and the projection of the satellite signal pierce points at 250 km, chosen as effective ionospheric altitude for TEC measurements, are shown in Figs. 1 and 2. Since there is always a station pair within 200 km of each other, small scale anomalies in TEC can be detected, that are shown in form of colour coded maps having a pixel size of 50 km by 50 km. In the following series of Figures an example of quiet conditions as well those showing mesoscale anomalies and their temporal development are shown (Figs. 3 to 5).

Since these anomalies are the result of medium size travelling ionospheric disturbances (MSTIDs) having horizontal wavelengths between 100 km and 1000 km and periods between 12 minutes and 1 hour, they are thought to be generated by atmospheric gravity waves (AGW) originating in the lower atmosphere, particularly since they occur at times when the ionosphere is not disturbed by "space weather". While ionospheric responses to energetic sources in the troposphere, such as tropical cyclones (hurricanes [2-4] and typhoons [5-7]) and even tsunamis [8] can be associated with their sources, this is not uniquely possible with MSTIDs [9] since AGW can travel large distances (up to 2000 km) from their tropospheric source before reaching ionospheric altitudes [10]. A possible association between a medium scale ionospheric anomaly and its tropospheric source is shown in Fig. 6. The ionospheric anomaly occurring over the Balkans seems to be associated with an upper troposphere funnel system connected with the occlusion of an extra-tropical cyclone in the lower troposphere. Observations with the Japanese MU- (middle and upper atmosphere-) Radar have shown that such an event is a possible source of AGW activity [11]. Thus the "birth" and "death" of extra-tropical cyclones may possibly be sources of short lived MSTID's in the ionosphere.

Statistics of the occurrence of mesoscale ionospheric anomalies for the first five years of our GPS observations, corresponding to an elevated level of solar activity, show that on 20 % to 25 % of the observation days ionospheric anomalies not associated with space weather are present, with 2/3 occurring during the "winter months" (November to March) [12]. The GPS range errors associated with these anomalies are in the decimetre to meter range. Because of the rising concern with high precision navigation it seemed worthwhile to extend the statistics of their occurrence into the second half of our 11 year observation period.

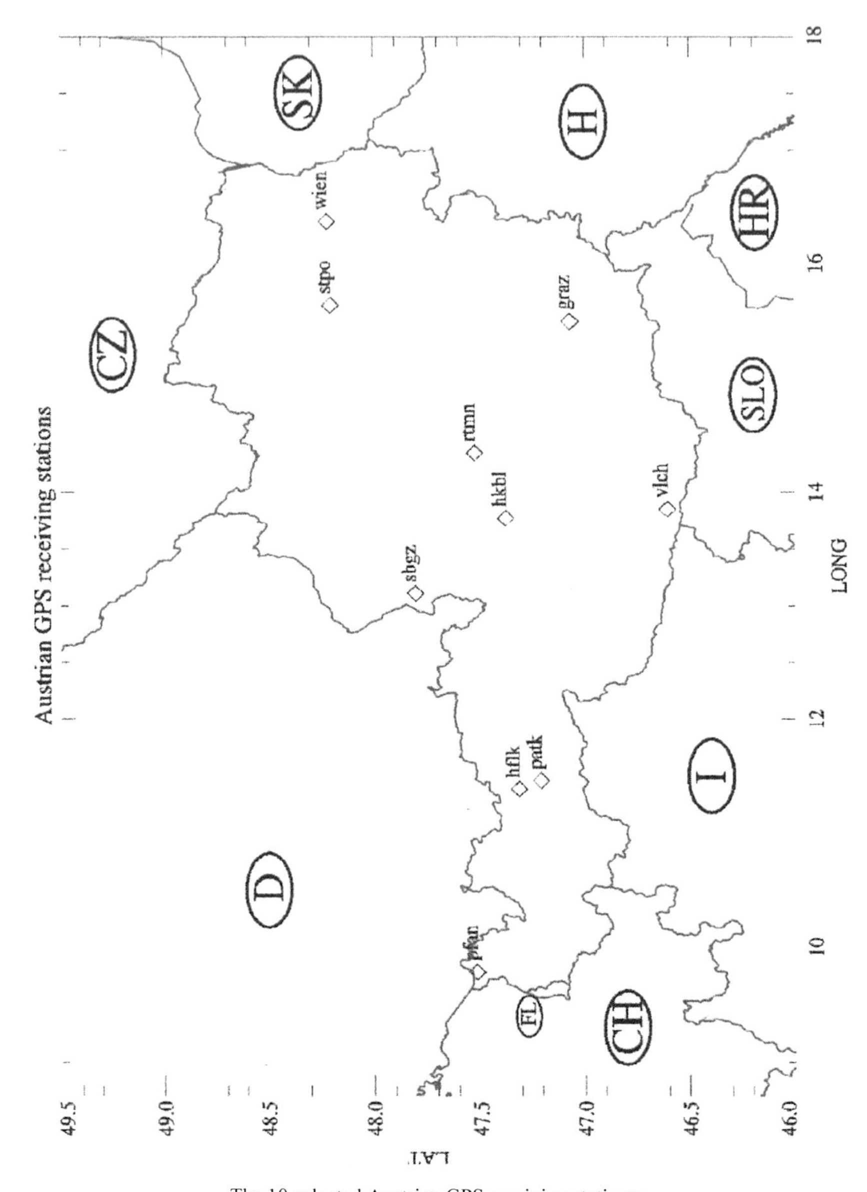

The 10 selected Austrian GPS receiving stations:

graz: Graz – Lustbühel (Styria)
hflk: Hafelekar (Tyrol)
hkbl: Hauser Kaibling (Styria)
patk: Patscherkofel (Tyrol)
pfan: Pfänder (Vorarlberg)
rtmn: Rottenmann (Styria)
sbgz: Salzburg (Salzburg Province)
stpo: St. Pölten (Lower Austria)
vlch: Villach (Carinthia)
wien: Wien (Vienna)

Figure 1: Austrian GPS dual frequency stations used for studying MSTIDs

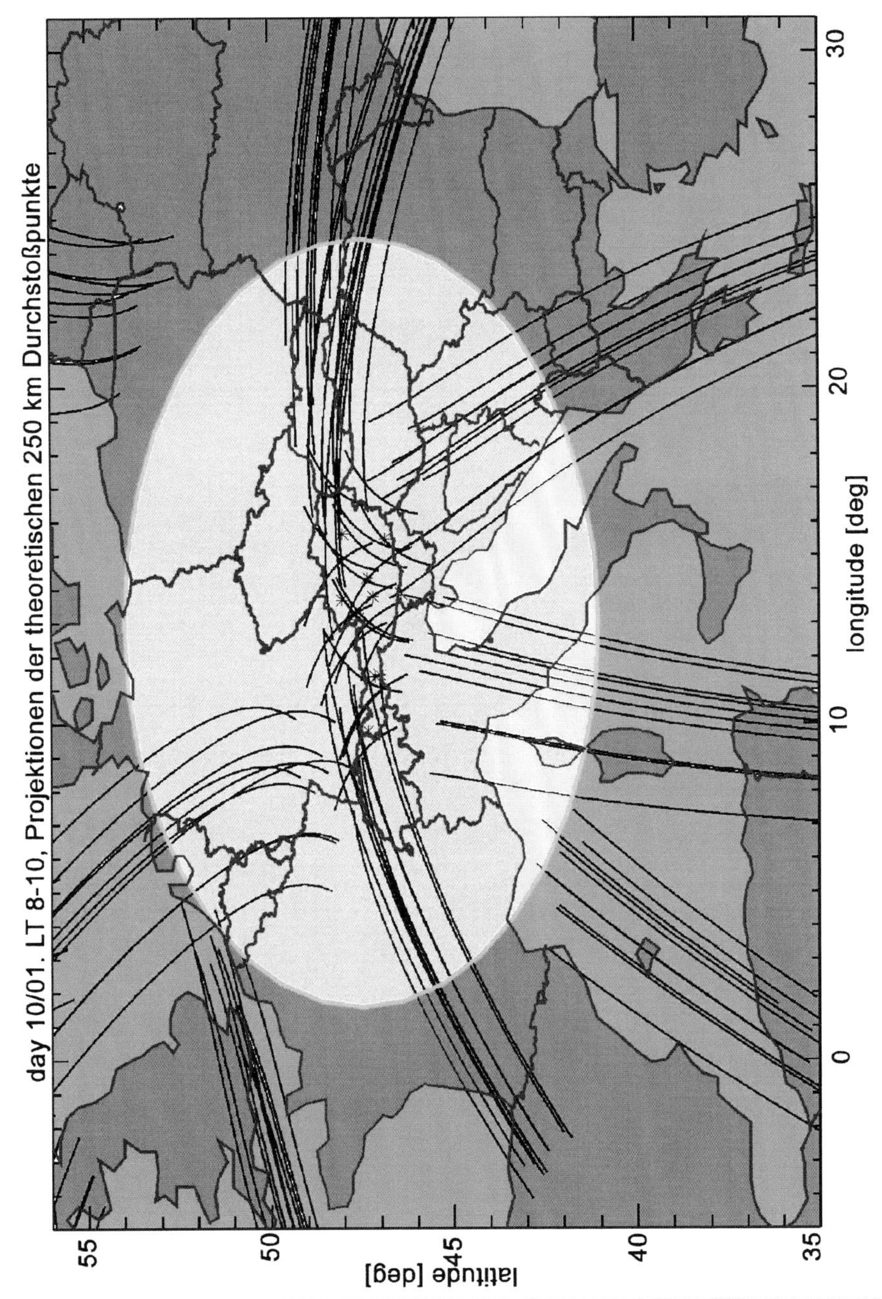

Figure 2: GPS signal pierce point projection at 250 km altitude as effective ionospheric altitude (oval shows approximate area of coverage by Austrian station net)

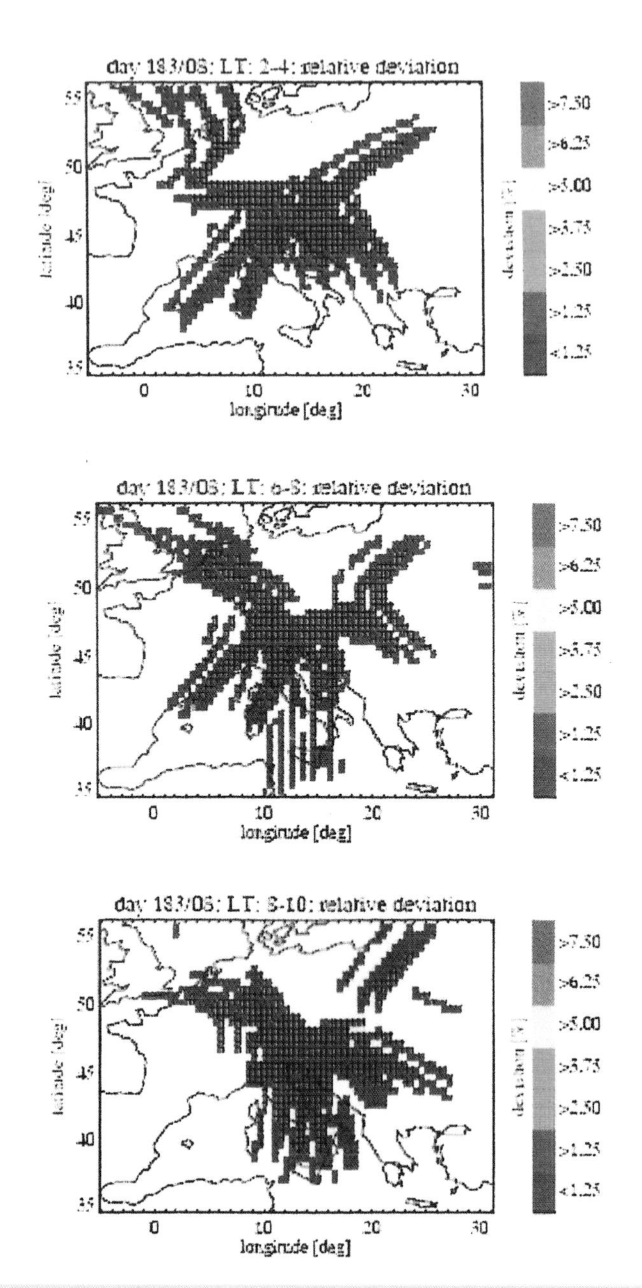

Figure 3: Colour coded pixel map indicating departures from background ionosphere (Dark blue approx. 1 %) July 7, 2008; UT 02:00 – 16:00

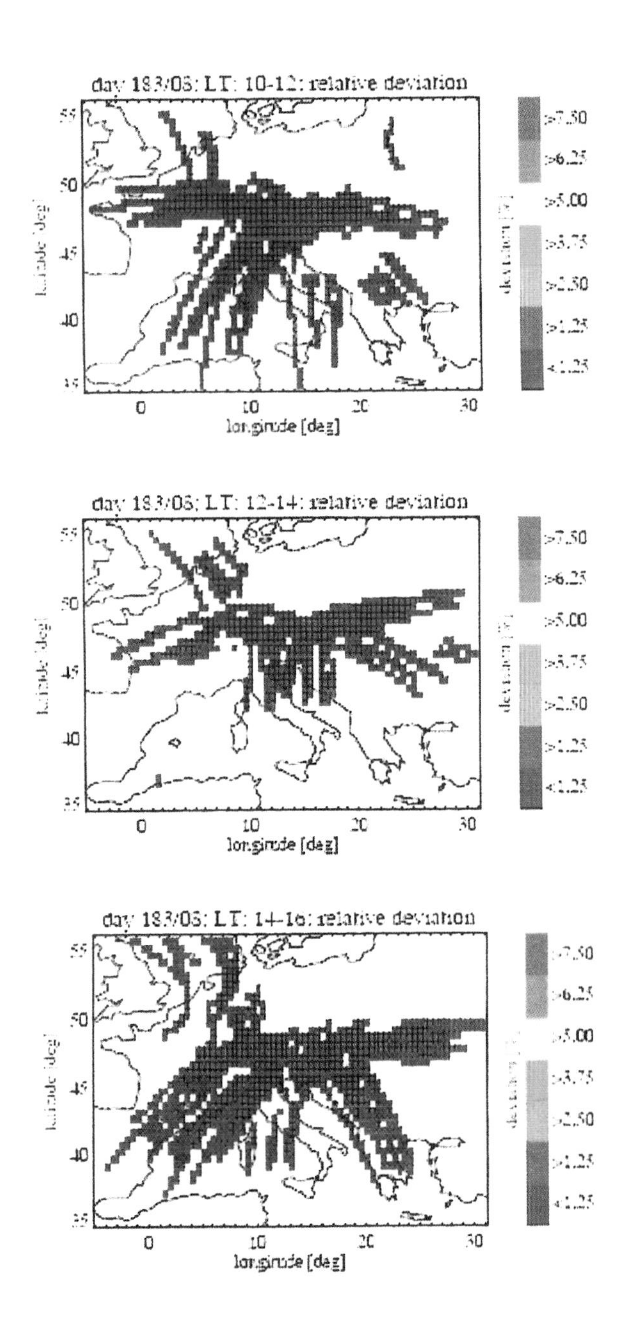

July 7, 2008; UT 02:00 – 16:00

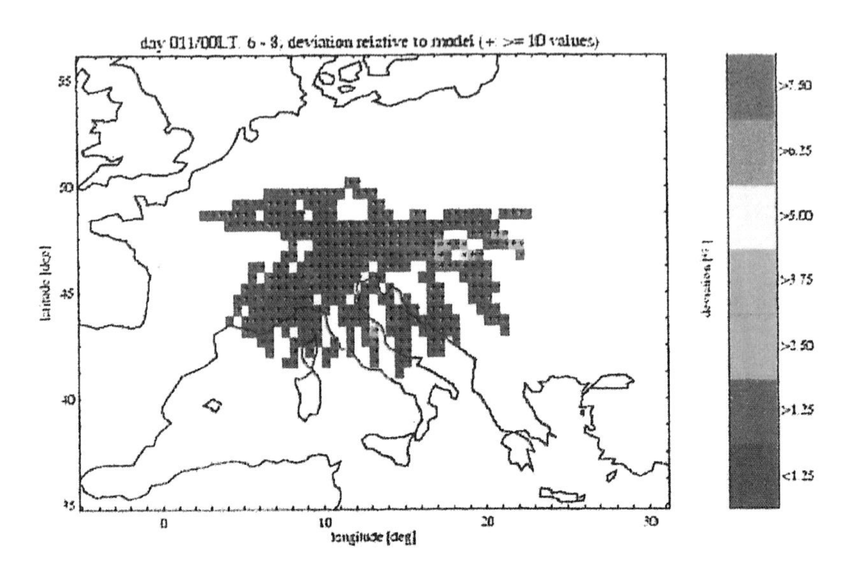

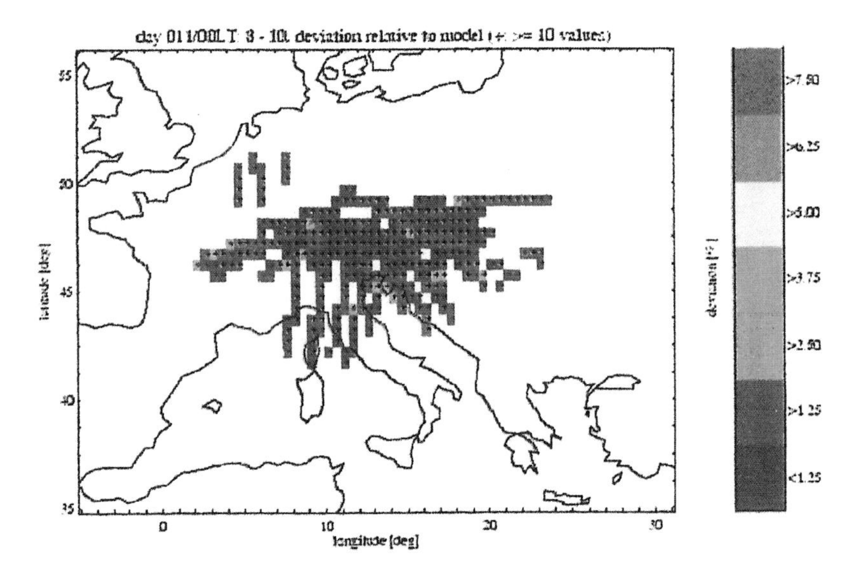

Figure 4: Development of an anomaly lasting about 2 hours (solar maximum); Range Error (RE) 1.4 m January 11, 2000; UT 06:00 – 14:00

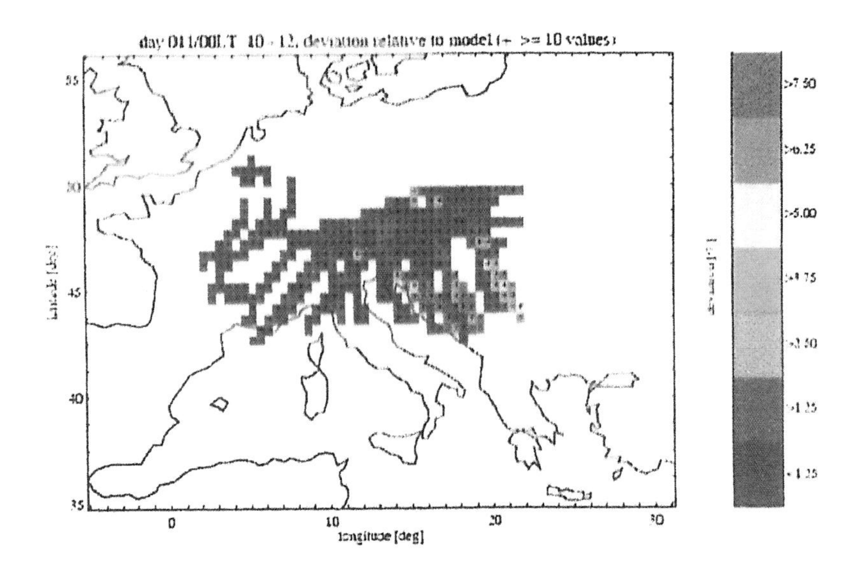

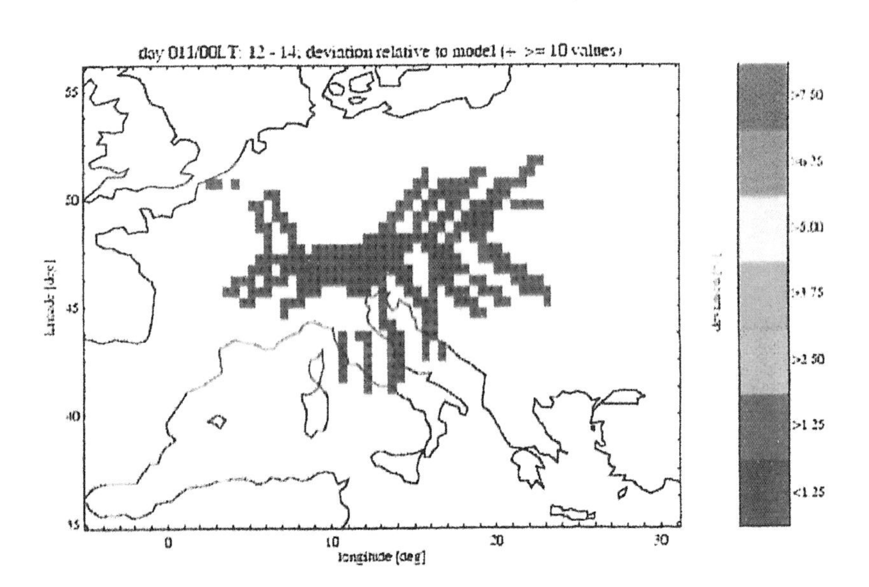

January 11, 2000; UT 06:00 – 14:00

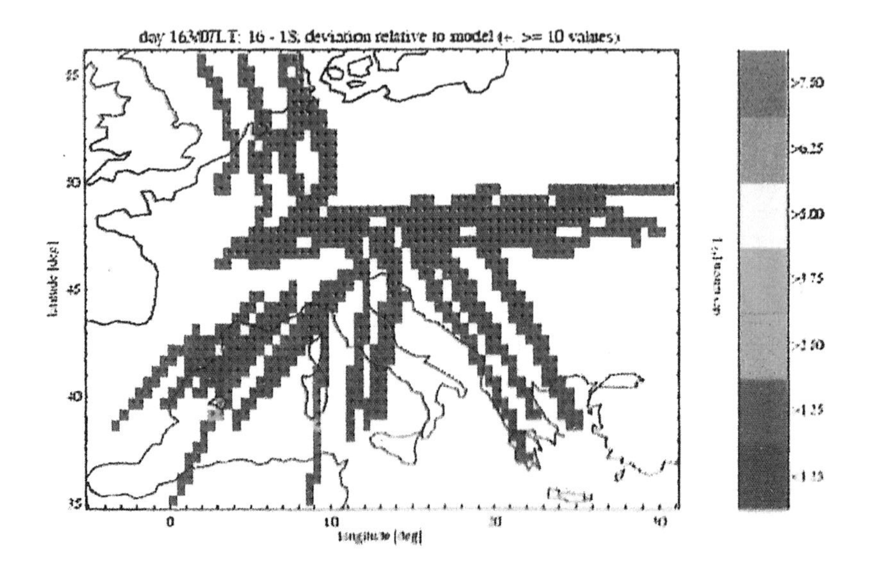

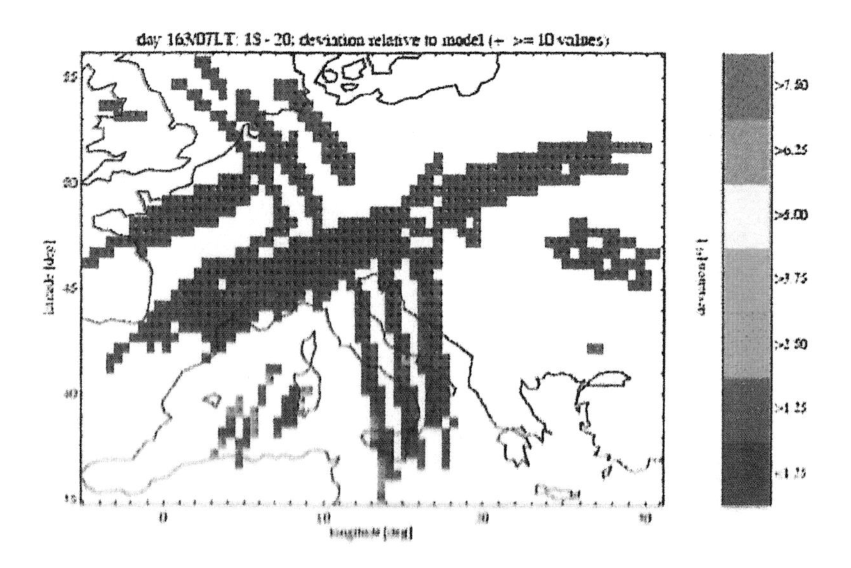

Figure 5: Anomaly near Gulf of Genoa (solar minimum) not included in statistics (RE 37 cm) July 12, 2007; UT 16:00 – 24:00

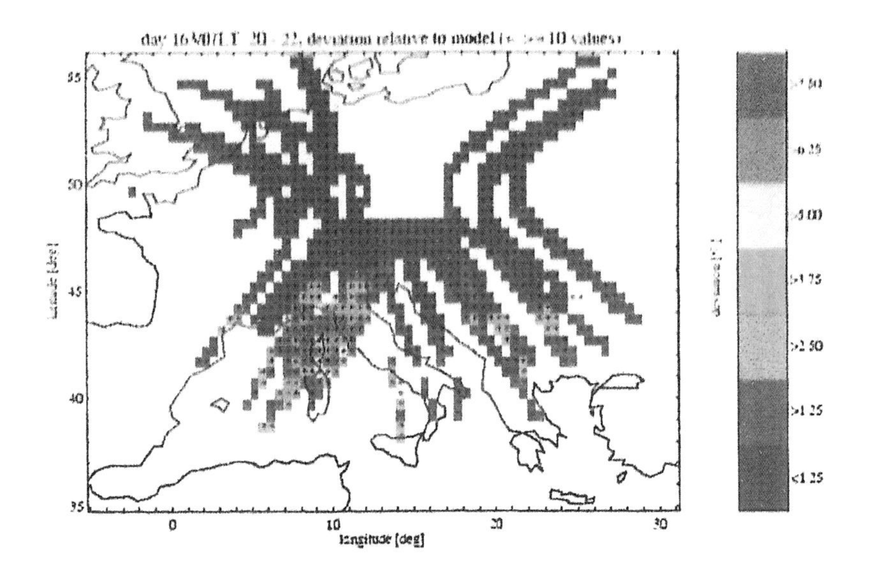

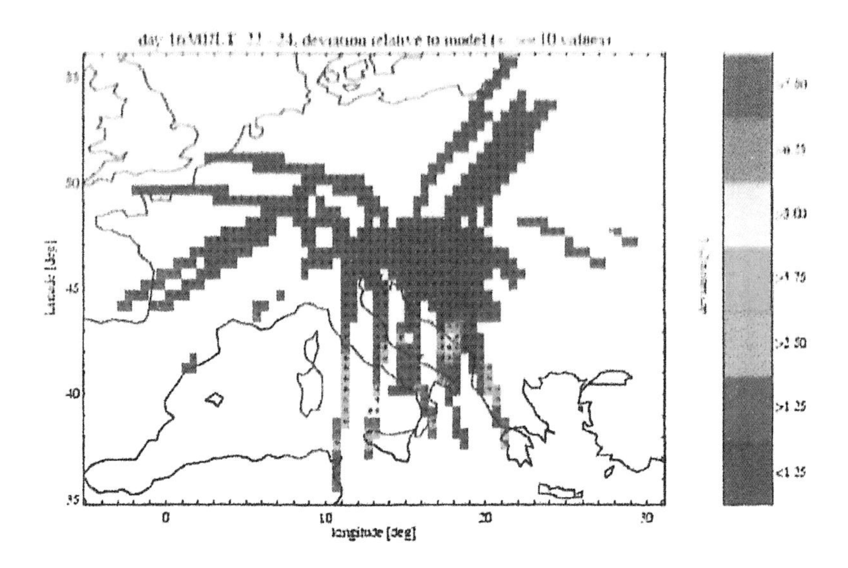

July 12, 2007; UT 16:00 – 24:00

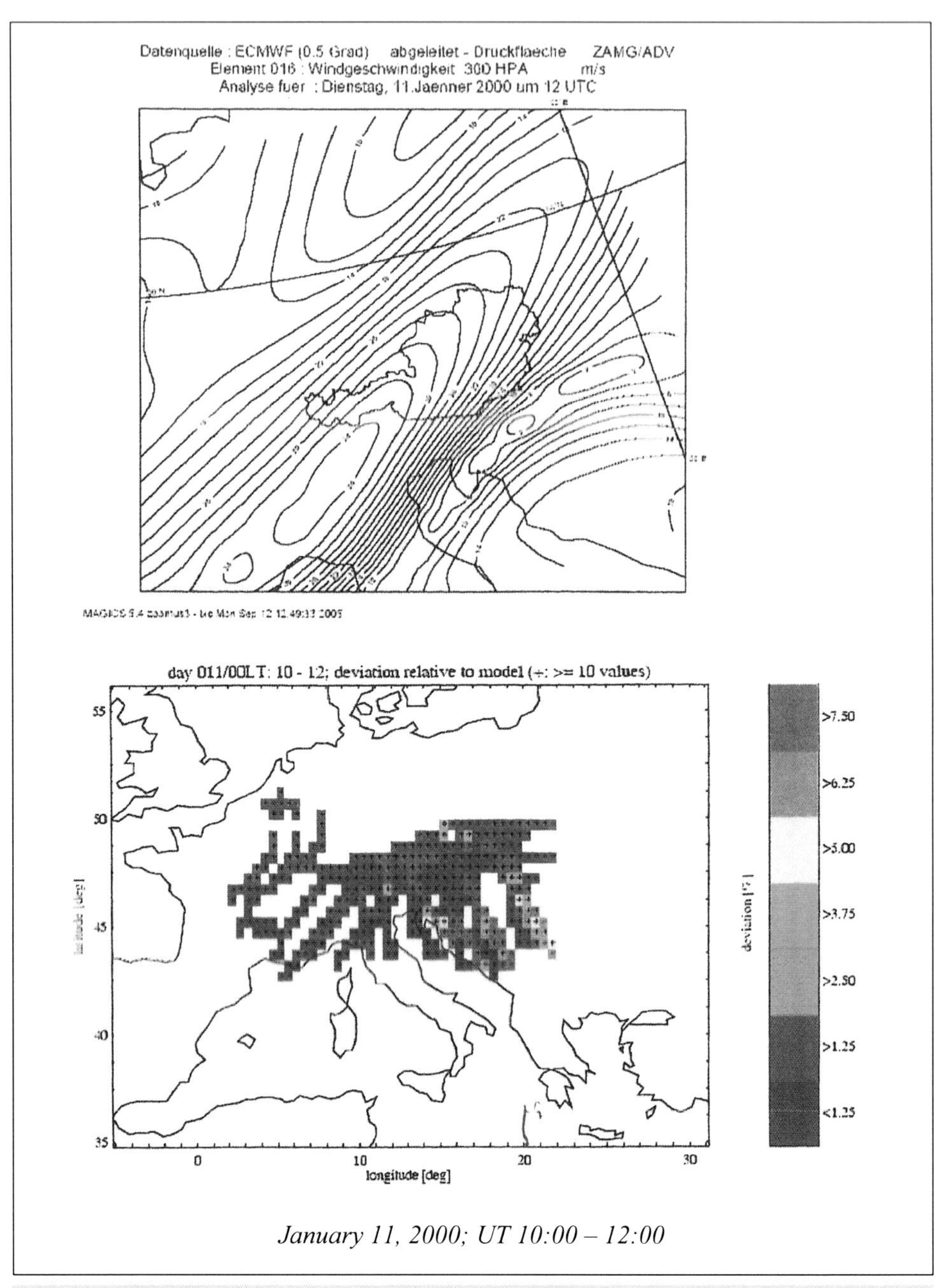

January 11, 2000; UT 10:00 – 12:00

Figure 6: Association between anomaly over Balkans (same as Fig. 4) and upper troposphere funnel system

Luckily enough, systematic GPS observations employing the Austrian net of 10 two-frequency GPS stations, whose location is known with high precision, started in the late nineties of last century, a period with elevated solar activity level and thus well observable effects in the respective data sets.

2 Statistics of Range Errors >0.5 m caused by Ionospheric Anomalies not associated with Space Weather

The statistics of range errors encompassing the entire period of GPS observations with the Austrian network produced some admittedly unexpected results. While the early period from 2000 to 2005 had shown a frequency of events, decreasing from about 25 % to less than 20 %, with threshold range errors less than a few decimetres, the new statistics with the higher threshold of 5 decimetre showed a complete lack of events beginning in 2006 as shown in Fig. 7 together with a plot of Relative Sunspot Numbers R_I and a similar distribution for range errors >1 m in Fig. 8. Thus, these range errors (shown in bi-monthly "bins") are obviously related to solar activity. Since the sources of these errors are thought to reside in the lower atmosphere and are propagated by atmospheric gravity waves to ionospheric F region levels, it is difficult to ascribe the statistics to tropospheric sources that must be independent of solar activity. The relationship obviously must be the result of changes in the ionosphere and thermosphere due to their dependence on solar EUV irradiance in the wavelength range from 1 nm to 100 nm as apparent in the comprehensive analysis of Total Electron Content climatology [13]. Fig. 9, reproduced from this publication, provides an impressive overview of the variability of the relevant parameters during solar cycle 23; ([Relative] Sunspot number R_I: $R_I = f + 10 \cdot g$; f: number of sunspots; g: number of sunspot groups).

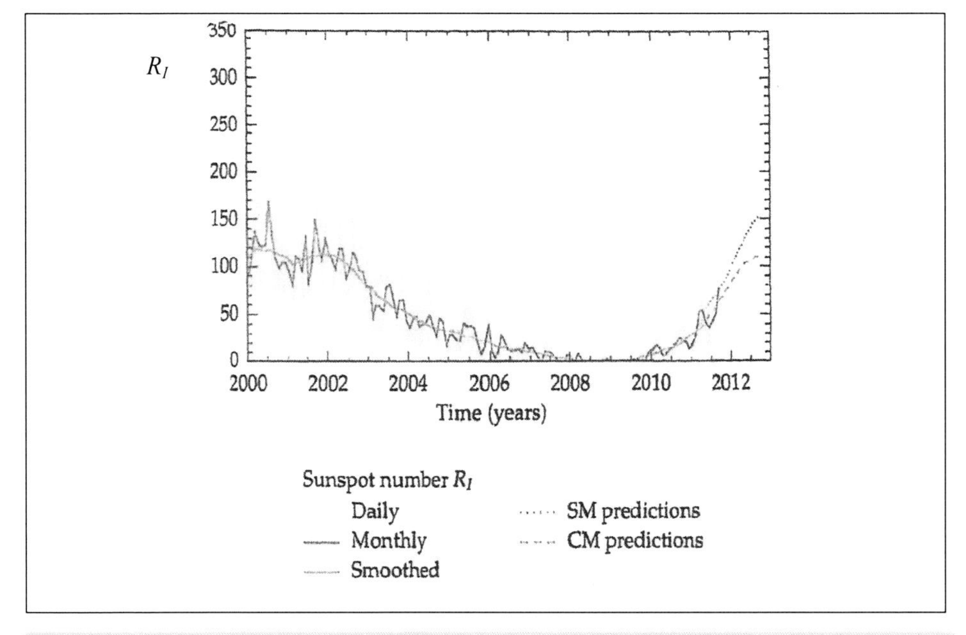

Figure 7a: Solar cycle 23 sunspot numbers R_I

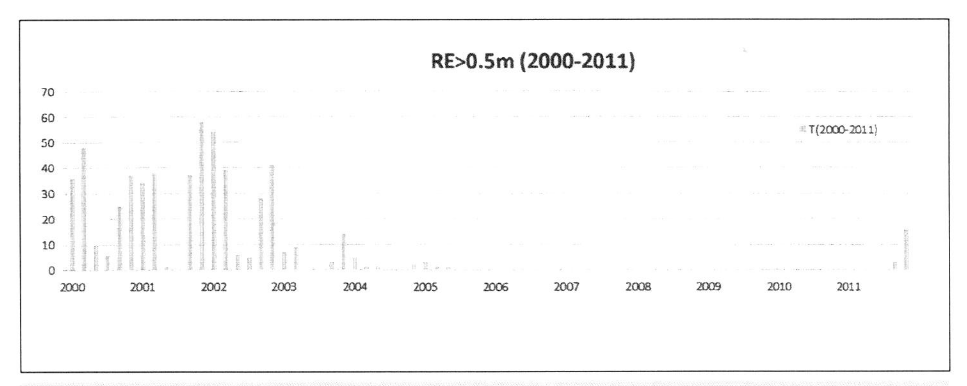

Figure 7b: Solar cycle 23 appearance statistics of Range Errors >0.5 m

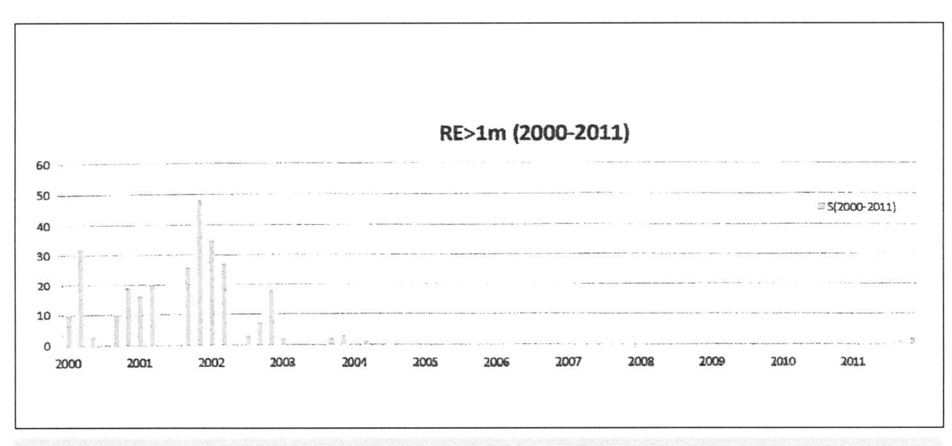

Figure 8: Solar cycle 23 appearance statistics of Range Errors >1 m

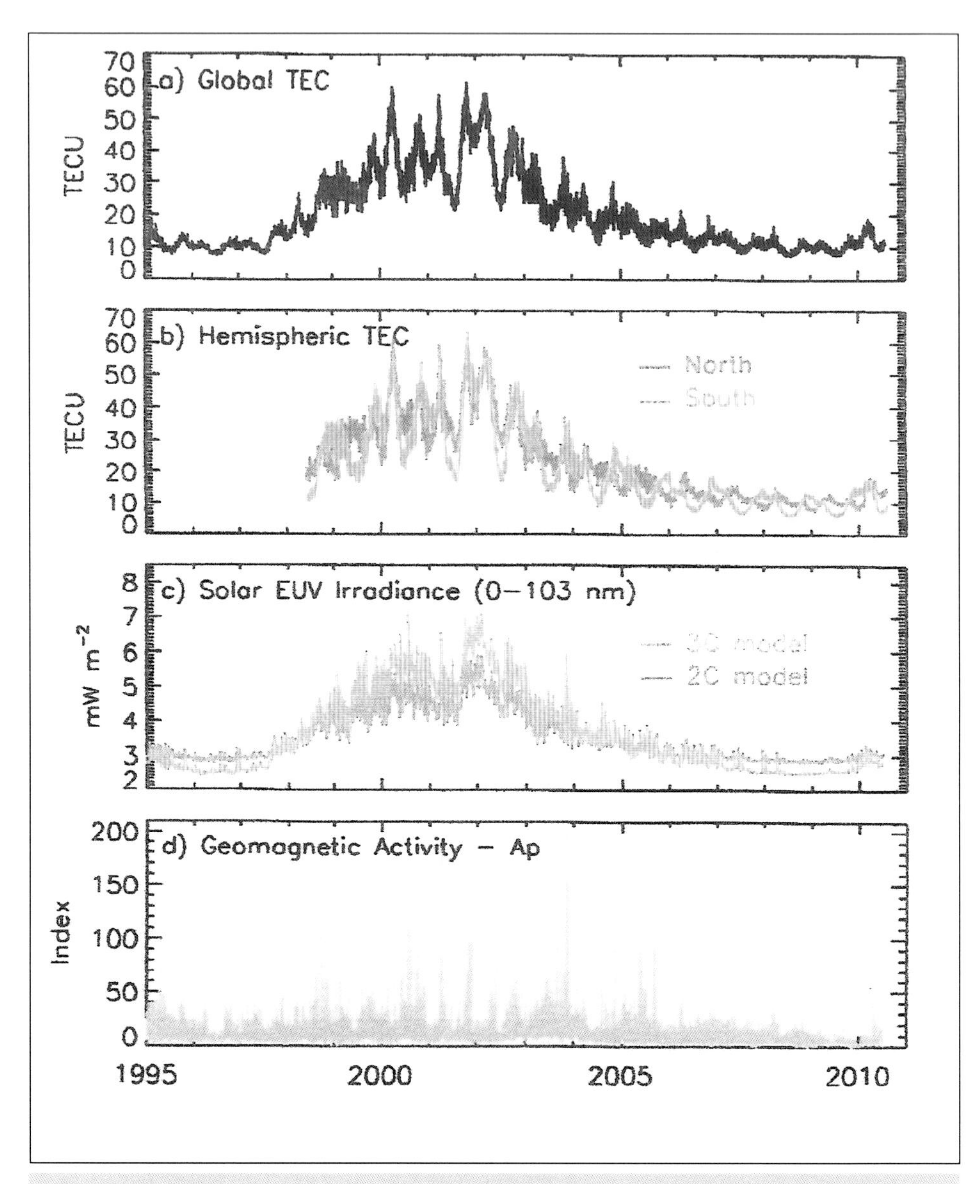

Figure 9: Total Electron Content (TEC), solar EUV irradiance and geomagnetic activity for the period from 1995 to 2010 (Reproduced from Reference [13] with permission of AGU)

3 Total Electron Content (TEC) as non-linear Function of Solar EUV Irradiance

In spite of observational evidence that TEC is a highly sensitive indicator of solar activity, there seems to be considerable reluctance to accept a non-linear relationship between them. Even the most comprehensive analysis of TEC climatology for solar cycle 23 [13] resorts to a linear model in accounting for the influence of solar and geomagnetic activity, semi-annual and annual oscillations and a secular trend.

In the following it will be shown that although F_2 region electron *density* reacts linearly to solar EUV irradiance, Total Electron *Content* (TEC) does not.

It is well known that the ionospheric F_2 region is the major contributor to TEC. In the following arguments for a non-linear relationship between TEC and solar variability, we shall apply accepted F_2 region theory [14] as summarized in Fig. 10.

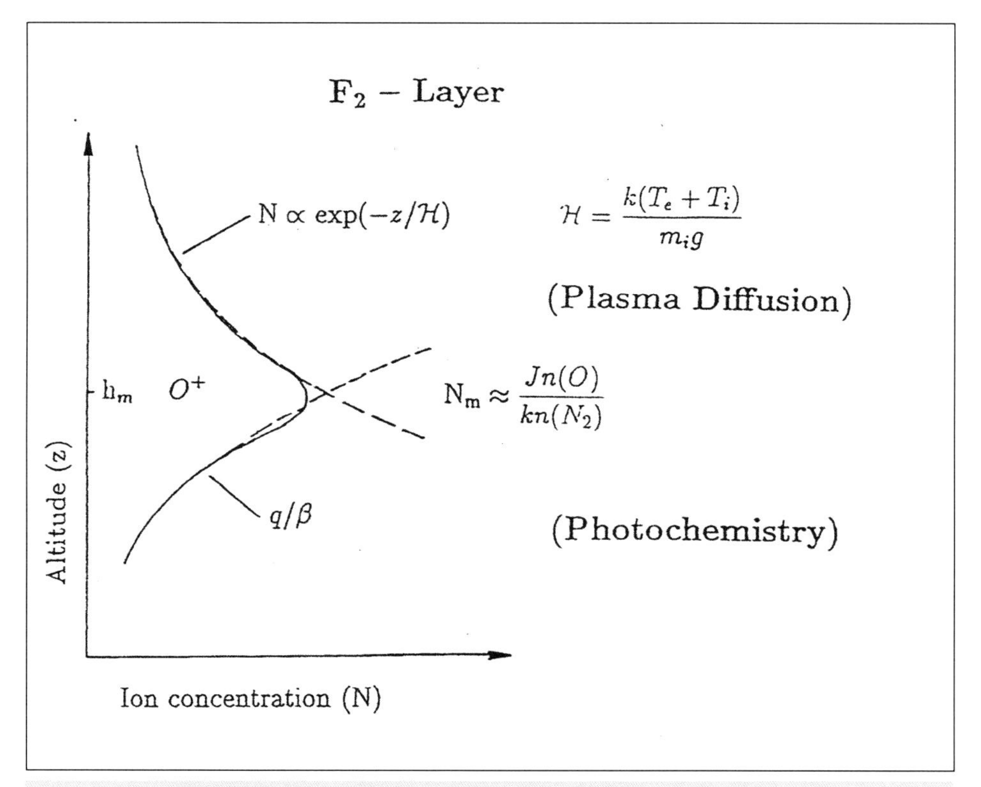

Figure 10: Summary of F_2 region formation theory (Reproduced from Reference [14] with kind permission)

The formation of the F_2 region is the result of two processes: photochemical equilibrium below the peak, and, plasma diffusion above, with N_m occurring at $h_m \approx 300$ km, where the time constants for the two processes equalize.

In contrast to a CHAPMAN layer, where the electron density maximum coincides with the ion pair production maximum, occurring at unit optical depth of the ionising radiation and where its N_m results from an equilibrium between ion pair production and recombination according to a quadratic loss law, the maximum of the F_2 region lies well above its ion production peak residing in the F_1 region.

Since the principal ion of the F_2 region is O^+, (radiative) recombination is much slower than an ion-molecule reaction involving N_2, the electron density is the result of a linear loss law, and thus is linearly dependent on solar ionising radiation at 91.1 nm. (In Fig. 10, J represents the ionisation frequency of atomic oxygen).

Since both, n(O), the concentration of the ionisable constituent and $n(N_2)$, involved in the chemical loss are exponentially *decreasing* with height according to their respective scale height, their ratio leads to an exponentially *increasing* behaviour with height of N with a scale height corresponding to an effective mass resulting from the mass difference of the two constituents N_2 (= 28 AMU) and O (= 16 AMU), AMU: Atomic Mass Unit, thus effective mass 12 AMU, $m_{(12)}$, i.e.

$$\mathbf{H = k \cdot T / m_{(12)} \cdot g} \qquad (1)$$

The ratio of O to N_2 is also responsible for the so called winter anomaly of the F_2 region, since the lighter oxygen is transported from the warmer summer to the winter hemisphere, thus increasing N_m in winter.

The electron density distribution above the F_2 peak on the other hand shows an exponentially decreasing function with a scale height resulting from plasma diffusion under gravity and thus depending on both, the temperature T of the electrons e, T_e, and ions i, T_i, where T_e can in principle be higher than the temperature T_i of ions and T_n of neutrals, when the condition

$$\mathbf{T_e > T_i \cong T_n} \qquad (2)$$

would generally hold. Under thermal equilibrium the scale height of an ion i with mass m_i would be twice that of a neutral of the same mass. If thermal equilibrium does not prevail, this factor could be greater than 2, viz. $(T_e / T_i + 1)$.

For lack of an analytical expression for the electron density distribution of the F_2 region, we can approximate TEC

$$\mathbf{TEC = \int N(z) \cdot dz} \tag{3}$$

as the sum of the electron content below and that above the F_2 peak occurring at h_m. Using a height parameter z

$$\mathbf{z = h - h_m} \tag{4}$$

with h_m as the reference level, integration with proper accounting for the lower and upper boundary values of the exponential electron density distribution above and below the peak, yields for the topside content N_m times $H(O^+)$ and for the bottomside content N_m times $H(m_{(12)})$, accordingly the ratio of topside to bottom side content is 1.5. This ratio is much too low when the results of early determination of ionospheric electron content by moon radar are considered that implied a ratio of at least 3 [15, 16]. Thus the scale height for the topside ionosphere with O^+ as the principal ion seems to be too small. In order to satisfy the requirement that the topside contribution to total content be the major one, a doubling of this scale height is required. The consequence of this doubling is a reduction in the ion mass by a factor of two. Thus instead of the mass of oxygen ions (16), the effective mass for the new scale height is 8. This implies a transition from oxygen ions to the lighter ion constituents such as He^+ and H^+ as illustrated in Fig. 11.

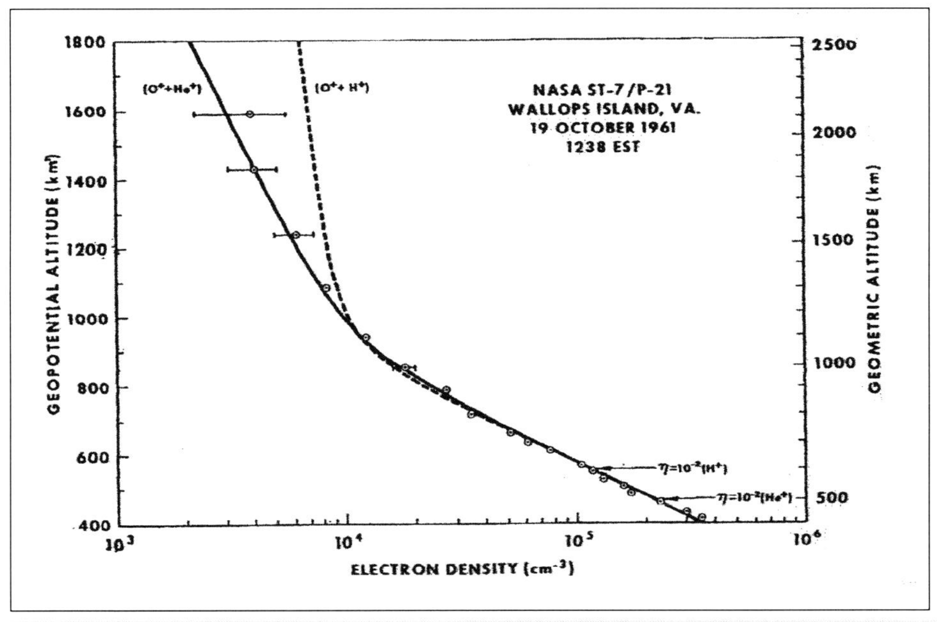

***Figure 11:** Electron density profile from FARADAY rotation measurements at 73.6 MHz and theoretical profiles for binary ion-mixtures O^+ with He^+ and H^+; (Reproduced from Reference [17] with kind permission)*

Although most of the electron density *profiles* of the topside ionosphere are limited to the height of the topside sounder satellites, i.e. about 1000 km, an early sounding rocket measurement (Fig. 11) had provided an electron density profile up to an altitude of about 2500 km that clearly shows a transition of the principal F region ion O^+ to lighter ions [17]. This profile is illustrated in Fig.11 which also shows theoretical distributions for binary ion mixtures in diffusive equilibrium according to an expression derived by the author [18].

Based on the above discussion we can represent TEC by the sum of bottomside and topside contents as

$$\mathbf{TEC \approx N_m \cdot (H_{topside}/3 + H_{topside}) = N_m \cdot (4/3) \cdot H_{topside} \sim N_m \cdot H_{eff.}} \qquad (5)$$

attesting to the importance of the topside ionosphere for TEC.

Regarding the functional dependence of TEC on solar EUV irradiance we can see that both factors have a dependence on I; (I: solar EUV irradiance from 1 nm to 100 nm): N_m through ionising radiation and $H_{topside}$ through thermospheric heat production by solar EUV energy flux (irradiance). The characteristic temperature T for the F region is also called exospheric temperature, since this asymptotically constant temperature in the thermosphere, due to heat input from the sun and downward heat transport by thermal conduction, is also representative for the base of the exosphere at an altitude of about 500 km, from where atmospheric escape can take place. Because of the role of heat conduction, this temperature was found to depend on the exponent s of the heat conduction's temperature dependence according to

$$\mathbf{T \sim I^{(1/s)}} \qquad [14, 19] \qquad (6)$$

Since $\mathbf{s = 0.71}$ for atomic oxygen, the principal constituent of the thermosphere,

$$\mathbf{T \sim I^{1.4}} \qquad (7)$$

Thus

$$\mathbf{TEC \sim I \cdot I^{1.4} = I^{2.4}} \qquad (8)$$

and

$$\mathbf{\Delta TEC/TEC = 2.4 \cdot \Delta I/I} \qquad (9)$$

This amplification factor allows for an almost five fold increase of TEC from solar minimum to solar maximum, in good agreement with observations [13].

4 Conclusions

The strong solar cycle dependence of appearance frequency of GPS range errors due to mesoscale ionospheric anomalies (MSTIDs) seems to be related to the non-linear response of TEC to solar EUV irradiance, since error size depends on TEC, but also on the propagation characteristics of AGW that require higher upper atmosphere temperatures (Solar max.) to reach ionospheric altitudes [20].

Acknowledgements

The author gratefully acknowledges the important contributions of M. RIEGER in maintaining the GPS observations data base and identifying MSTID activity.

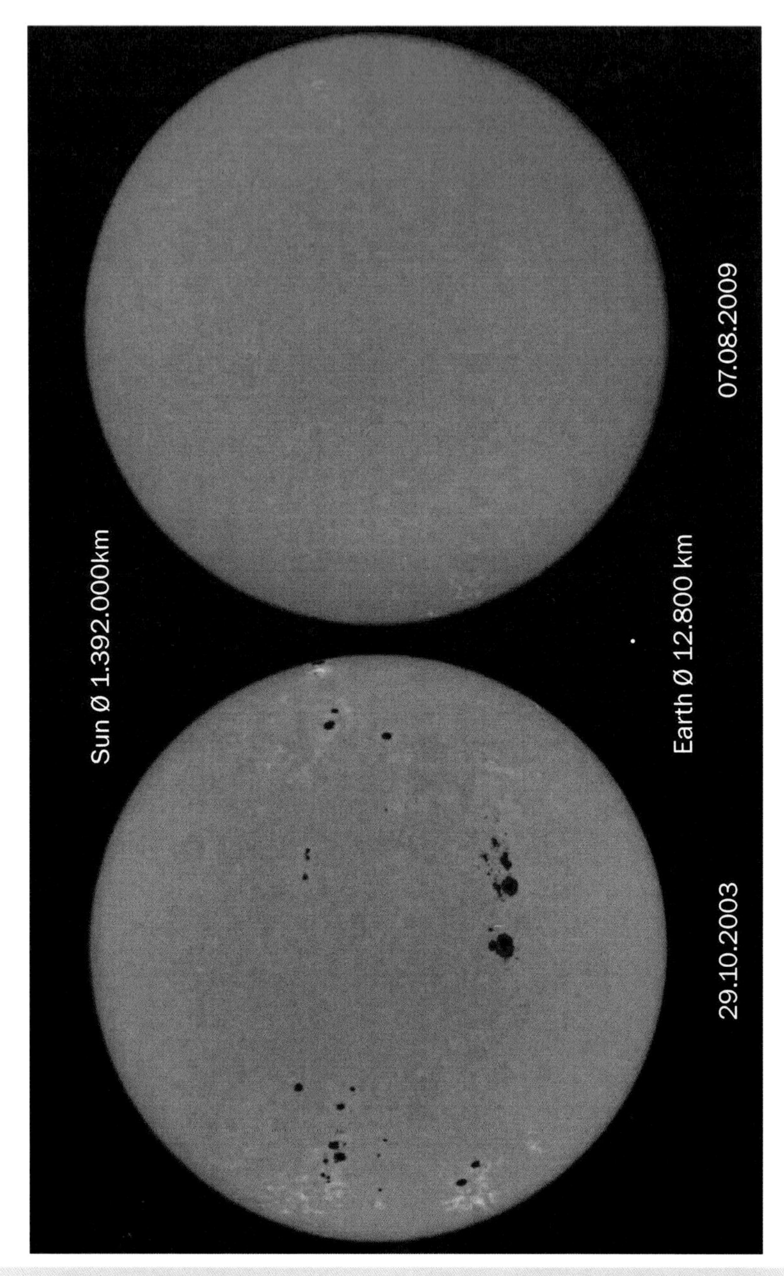

Figure 12: Sunspot cycle 23: Solar images dating 29.10.2003 (spots) and 07.08.2009 (spot free); (with kind permission by KANZELHÖHE Observatory, CARINTHIA; KARL-FRANZENS-University, GRAZ, AUSTRIA

5 References

[1] **Bauer, S. J.:** Die Abhängigkeit der Nachrichtenübertragung, Ortung und Navigation von der Ionosphäre, Projektbericht 4, 88 pp. Austrian Academy of Sciences Press, 2002 (ISBN 3-7001-3140-2).

[2] **Bauer, S. J.:** An apparent ionospheric response to the passage of hurricanes, J. Geophys. Res. 63, 265-269, 1958.

[3] **Hung, R. J. and Kuo, J. P.:** Ionospheric observation of Gravity waves associated with hurricane Eloise, J. Geophys. Res. 45, 67-80, 1978.

[4] **Bishop, R., Aponte, N., Earle, G. D., Sulzer, M., Larsen, M. F. and Peng, G.:** Arecibo observations of ionospheric perturbations associated with the passage of tropical storm Odette, J. Geophys. Res. 111, A11320, 2006.

[5] **Shen, C. S.:** The correlation between the typhoon and the $f_0(F_2)$ of the ionosphere, Chin. J. Space Science, 2 (4), 335-340, 1982.

[6] **Tian Mao, JingSong Wang, GuangLin Yang, JinSong Ping and YuCheng Suo:** Effects of typhoon Matsa on ionospheric TEC, Chinese Science Bulletin, Vol. 55 (8), 712-717, 2010.

[7] **Rice, D. D., Sojka, J. J., Eccles, J. V. and Schunk, R. W.:** Typhoon Melor and ionospheric weather in the Asian sector: a case Study, Radio Science, Vol. 47, RSOLO5, 9 pp., 2012.

[8] **Liu, J.-Y., Chen, C.-H., Lin, C.-H., Tsai, H.-F., Chen, C.-H. and Kamogawa, M.:** Ionospheric disturbance created by the 11 March 2011 M9.0 Tohoku earthquake, J. Geophys. Res., Vol. 116, A06319, 5 pp., 2011.

[9] **Rieger, M. and Leitinger, R.:** The Ground Weather Connection – Medium Scale TIDs in the Ionosphere: SJB 75 – Festschrift for Prof. S. J. Bauer (H. O. Rucker and R. Leitinger, eds.), 111-121, Austrian Academy of Sciences / University of Graz), 2005 (ISBN 3-902081-01-5).

[10] **Vadas, S. L.:** Horizontal and vertical propagation of gravity waves in the thermosphere from lower atmosphere & thermosphere sources, J. Geophys. Res., Vol. 112, AO6305, 2007.

[11] **Fukao, S., Yamanaka, M. D., Matsumoto, H., Sato, T., Tsuda, T. and Kato, S.:** Wind fluctuations near a cold vortex-tropopause funnel system observed by MU radar, PAGEOPH. Vol. 130 (2/3), 463, 1989.

[12] **Rieger, M., Leitinger, R. and Bauer, S. J.:** Mesoscale ionospheric anomalies not associated with space weather, Radio Science, Vol. 141, RS6S10, 11 pp., 2006.

[13] **Lean, J. L., Meier, R. R., Picone, J. M. and Emmert, J. T.:** Ionospheric total content: global and hemispheric climatology, J. Geophys. Res., Vol. 116, A10318 (18 pages), 2011.

[14] **Bauer, S. J. and Lammer H.:** Planetary Aeronomy, 207 pp, Springer Verlag, 2004 (ISBN 1610-1677 and ISBN 3-540-21472-0).

[15] **Evans, J. V.:** The electron content of the ionosphere, J. Atm. Terr. Physics, 11, 259-271, 1957.

[16] **Bauer, S. J. and Daniels, F. B.:** Measurement of ionospheric electron content by the Lunar radio technique, J. Geophys. Res., Vol. 64 (10), 1371-1376, 1959.

[17] **Bauer, S. J. and Jackson, J. E.:** Rocket measurement of the electron density distribution in the topside ionosphere, J. Geophys. Res., Vol. 67 (4), 1675-1677, 1962.

[18] **Bauer, S. J.:** The electron density distribution above the F_2 peak and associated atmospheric parameters, J. Atm. Sci. Vol. 19, 17-19, 1962.

[19] **Bauer S. J.:** Solar cycle variations of planetary exospheric temperatures, Nature (Phys. Sci.), Vol. 232, 101-102, 1971.

[20] **Fritts, D. C. and Vadas, S. L.:** Gravity wave penetration into the thermosphere: sensitivity to solar activity and mean wind, Ann. Geophys., 26, 3841-3861, 2008.

6 Figure Captions

Em. o. Univ.-Prof. Dr. Siegfried J. Bauer

Photo: Karl Acham

Siegfried J. Bauer was born at Klagenfurt, Austria in 1930. From 1948 to 1953 he studied physics, geophysics and meteorology at the University of Graz, receiving a doctorate (Phil. Dr.) with a dissertation in ionospheric research under O. Burkard.

Soon afterwards he came to the United States through the military "Project Paperclip" to work at the U. S. Army Signal Research and Development Laboratory, Fort Monmouth, N. J., where he remained until 1960. There his research was concerned with atmospheric effects on radio propagation, weather radar and the ionospheric response to hurricanes and surface atomic explosions. During the last three years he conducted measurements of the ionospheric electron content with the DIANA moon radar.
In 1961 he joined the newly-established NASA Goddard Space Flight Center, Greenbelt, MD, where he remained for two decades. There his personal research concentrated on rocket and satellite experiments of the topside ionosphere and associated theoretical studies and later also on the ionospheres of Venus and Mars. During his NASA period he also served as Head, Ionosphere and Radio Physics Branch, Associate Chief, Laboratory for Planetary Atmospheres and last as Associate Director of Sciences.

In 1981 he returned to his native Austria upon his appointment to the Chair of Meteorology and Geophysics at the University of Graz, a position once held by Alfred Wegener of continental drift fame; in addition he was a Department Head at the Space Research Institute of the Austrian Academy of Sciences until becoming Emeritus Professor in the fall of 1998.

His personal research in Austria focused on planetary atmospheres and ionospheres on account of his association with the American Pioneer Venus and Mars Global Surveyor missions and the European HUYGENS Probe that landed on TITAN. In addition to numerous publications, he is the author of the classic monograph "Physics of Planetary Ionospheres" that also appeared in Russian and Japanese, and with H. Lammer of "Planetary Aeronomy".

He holds membership in the Austrian Academy of Sciences, the Academia Europaea and the International Academy of Astronautics. He also is a Fellow of AAAS and AGU and Honorary Fellow of the Royal Astronomical Society and recipient of the NASA Exceptional Scientific Achievement Medal and the David Bates Medal of the European Geophysical Society (now EGU).

e-mail: siegfried.bauer@uni-graz.at or: siegfried.bauer@oeaw.ac.at

VERLAG DER ÖSTERREICHISCHEN AKADEMIE DER WISSENSCHAFTEN
WIEN 2012

Folgende Publikationen sind inzwischen erschienen:

- **Projektbericht 1:**
  Elisabeth Lichtenberger: Geopolitische Lage und Transitfunktion Österreichs in Europa. Wien 1999.
- **Projektbericht 2:**
  Klaus-Dieter Schneiderbauer und Franz Weber (mit einem Beitrag von Wolfgang Pexa): Stoß- und Druckwellenausbreitung von Explosionen in Stollensystemen. Wien 1999.
- **Projektbericht 3:**
  Elisabeth Lichtenberger: Analysen zur Erreichbarkeit von Raum und Gesellschaft in Österreich. Wien 2001.
- **Projektbericht 4:**
  Siegfried J. Bauer (mit einem Beitrag von Alfred Vogel): Die Abhängigkeit der Nachrichtenübertragung, Ortung und Navigation von der Ionosphäre. Wien 2002.
- **Projektbericht 5:**
  Klaus-Dieter Schneiderbauer und Franz Weber (mit einem Beitrag von Alfred Vogel): Integrierte geophysikalische Messungen zur Vorbereitung und Auswertung von Großsprengversuchen am Erzberg/Steiermark. Wien 2003.
- **Projektbericht 6:**
  Georg Wick und Michael Knoflach: Kardiovaskuläre Risikofaktoren bei Stellungspflichtigen mit besonderem Augenmerk auf die Immunreaktion gegen Hitzeschockprotein 60. Wien 2004.
- **Projektbericht 7:**
  Hans Sünkel und Alfred Vogel (Hrsg.): Wissenschaft – Forschung – Landesverteidigung: 10 Jahre ÖAW – BMLV/LVAk. Wien 2005.

- **Projektbericht 8:**
  Andrea K. Riemer und Herbert Matis: Die Internationale Ordnung am Beginn des 21. Jahrhunderts. Eigenschaften, Akteure und Herausforderungen im Kontext sozialwissenschaftlicher Theoriebildung. Wien 2006.

- **Projektbericht 9:**
  Roman Lackner, Matthias Zeiml, David Leithner, Georg Ferner, Josef Eberhardsteiner und Herbert A. Mang: Feuerlastinduziertes Abplatzverhalten von Beton in Hohlraumbauten. Wien 2007.

- **Projektbericht 10:**
  Michael Kuhn, Astrid Lambrecht, Jakob Abermann, Gernot Patzelt und Günther Groß: Die österreichischen Gletscher 1998 und 1969, Flächen und Volumenänderungen. Wien 2008.

- **Projektbericht 11:**
  Hans Wallner, Alfred Vogel und Friedrich Firneis: Österreichische Akademie der Wissenschaften und Streitkräfte 1847 bis 2009 – Zusammenarbeit im Staatsinteresse. Wien 2009.

- **Projektbericht 12:**
  Andreas Stupka, Dietmar Franzisci und Raimund Schittenhelm: Von der Notwendigkeit der Militärwissenschaften. Wien 2010.

- **Projektbericht 13:**
  Guido Korlath: Zur Mobilität terrestrischer Plattformen. Wien 2011.

- **Projektbericht 14:**
  Siegfried J. Bauer: Solar Cycle Influence of GPS Range Errors from Mesoscale Ionospheric Anomalies (MSTIDs). Wien 2012.